1,000,000 Books

are available to read at

Forgotten Books

---

www.ForgottenBooks.com

Read online
Download PDF
Purchase in print

ISBN 978-0-365-68853-2
PIBN 11255296

This book is a reproduction of an important historical work. Forgotten Books uses state-of-the-art technology to digitally reconstruct the work, preserving the original format whilst repairing imperfections present in the aged copy. In rare cases, an imperfection in the original, such as a blemish or missing page, may be replicated in our edition. We do, however, repair the vast majority of imperfections successfully; any imperfections that remain are intentionally left to preserve the state of such historical works.

1 MONTH OF
FREE
READING

at

www.ForgottenBooks.com

By purchasing this book you are eligible for one month membership to ForgottenBooks.com, giving you unlimited access to our entire collection of over 1,000,000 titles via our web site and mobile apps.

To claim your free month visit:

www.forgottenbooks.com/free1255296

ATIONAL BUREAU OF STANDARDS REPORT

2498

AN ILLUSTRATION OF COMPUTATIONAL METHODS FOR THE
DETERMINATION OF THE PARAMETERS OF A CERTAIN
RANDOM PROCESS *

by

Eugene Lukacs

*The preparation of this paper was sponsored by the
U. S. Naval Ordnance Test Station, Inyokern.

NBS

U. S. DEPARTMENT OF COMMERCE
NATIONAL BUREAU OF STANDARDS

U. S. DEPARTMENT OF COMMERCE

Sinclair Weeks, *Secretary*

NATIONAL BUREAU OF STANDARDS

A. V. Astin, *Director*

THE NATIONAL BUREAU OF STANDARDS

The scope of activities of the National Bureau of Standards is suggested in the following listing of the divisions and sections engaged in technical work. In general, each section is engaged in specialized research, development, and engineering in the field indicated by its title. A brief description of the activities, and of the resultant reports and publications, appears on the inside of the back cover of this report.

Electricity. Resistance Measurements. Inductance and Capacitance. Electrical Instruments. Magnetic Measurements. Applied Electricity. Electrochemistry.

Optics and Metrology. Photometry and Colorimetry. Optical Instruments. Photographic Technology. Length. Gage.

Heat and Power. Temperature Measurements. Thermodynamics. Cryogenics. Engines and Lubrication. Engine Fuels. Cryogenic Engineering.

Atomic and Radiation Physics. Spectroscopy. Radiometry. Mass Spectrometry. Solid State Physics. Electron Physics. Atomic Physics. Neutron Measurements. Infrared Spectroscopy. Nuclear Physics. Radioactivity. X-Rays. Betatron. Nucleonic Instrumentation. Radiological Equipment. Atomic Energy Commission Instruments Branch.

Chemistry. Organic Coatings. Surface Chemistry. Organic Chemistry. Analytical Chemistry. Inorganic Chemistry. Electrodeposition. Gas Chemistry. Physical Chemistry. Thermochemistry. Spectrochemistry. Pure Substances.

Mechanics. Sound. Mechanical Instruments. Aerodynamics. Engineering Mechanics. Hydraulics. Mass. Capacity, Density, and Fluid Meters.

Organic and Fibrous Materials. Rubber. Textiles. Paper. Leather. Testing and Specifications. Polymer Structure. Organic Plastics. Dental Research.

Metallurgy. Thermal Metallurgy. Chemical Metallurgy. Mechanical Metallurgy. Corrosion.

Mineral Products. Porcelain and Pottery. Glass. Refractories. Enameled Metals. Concreting Materials. Constitution and Microstructure. Chemistry of Mineral Products.

Building Technology. Structural Engineering. Fire Protection. Heating and Air Conditioning. Floor, Roof, and Wall Coverings. Codes and Specifications.

Applied Mathematics. Numerical Analysis. Computation. Statistical Engineering. Machine Development.

Electronics. Engineering Electronics. Electron Tubes. Electronic Computers. Electronic Instrumentation.

Radio Propagation. Upper Atmosphere Research. Ionospheric Research. Regular Propagation Services. Frequency Utilization Research. Tropospheric Propagation Research. High Frequency Standards. Microwave Standards.

Ordnance Development. These three divisions are engaged in a broad program of research
Electromechanical Ordnance. and development in advanced ordnance. Activities include
Ordnance Electronics. basic and applied research, engineering, pilot production, field testing, and evaluation of a wide variety of ordnance matériel. Special skills and facilities of other NBS divisions also contribute to this program. The activity is sponsored by the Department of Defense.

Missile Development. Missile research and development: engineering, dynamics, intelligence, instrumentation, evaluation. Combustion in jet engines. These activities are sponsored by the Department of Defense.

● Office of Basic Instrumentation   ● Office of Weights and Measures.

UREAU OF STANDARDS REPORT

NBS REPORT

May 14, 1953   2498

ION OF COMPUTATIONAL METHODS FOR THE
ION OF THE PARAMETERS OF A CERTAIN
RANDOM PROCESS

by

Eugene Lukacs

An Illustration of Computational Methods for the
Determination of the Parameters of a Certain
Random Process*

by

Eugene Lukacs
National Bureau of Standards

1. **Introduction:** Statistical evaluation of the result of
measurements is called for whenever repeated measurements of
the same quantity are taken. However one encounters quite fre-
quently situations where the physical quantity observed varies
with time. We can then take a sequence of observations, but
due to the variability in time we can not say that they are
measurements of the same quantity. Moreover the situation can
be such that it is impossible to take simultaneously several
independent observations or to repeat the experiment under
identical conditions. As an example we mention the motion of
certain physical bodies which receive an initial impulse and
are then moving freely subject only to random influences.
The observed quantity would be in this case either the position or the velocity or the acceleration of the body. The
motion of a wide variety of missiles belongs into this class.
As another example we mention the decrease in thickness (or
weight) of a shoe sole or a tire due to the natural wear over
a period of time.

---

*The preparation of this paper was sponsored by the U. S. Naval
Ordnance Test Station, Inyokern.

In situations of this kind, the standard statistical techniques can not be applied in a natural manner. It seems
therefore desirable to find an appropriate probabilistic
model.

Although we can take observations only at discrete time
points, it seems quite natural to suppose that a random var-
iable (for instance the position of a moving body) is given at
each instant of time. It is customary to try to make as many
observations as possible, the temporal proximity of the observations as well as the nature of the physical phenomenon
will in most cases prevent the observations from being stoch-
astically independent. This is a situation which is rarely
studied in the theory of the statistical evaluation of measurements. However these considerations suggest strongly that a
stochastic process, depending on a continuous time parameter
would be the suitable probabilistic model for such a sequence
of observations.

A mathematically simple model is obtained if we assume that
the actual sequence of observations comes from a Wiener process with a certain mean value function. Such a model contains several parameters, the variance constant of the process
and the parameters introduced by the mean value function.
Procedures have been developed [2] [3] [4] for the estimation
of these parameters. The purpose of this paper is to give a

brief survey of the estimation procedures, the numerical aspects of obtaining these estimates will be emphasized. Finally an example of a Wiener process with mean value function is constructed by means of random numbers. The main purpose of the present paper is to illustrate the available statistical techniques by means of this artificial process.

2. The Wiener process

A stochastic process $x(t)$, depending on a continuous time parameter, is said to be a Wiener process if

(i) $x(t)$ is a process with independent increments and initial value $x(0) = 0$.

(ii) the increment $x(t+\tau) - x(t)$ is normally distributed with mean zero and variance $c\tau$, (where $c > 0$).

The constant c is called the variance constant of the process. The value $x(t)$ of the process can be written according to (i) as an increment, $x(t) = x(t) - x(0)$ so that we see from (ii) that the mean value function $Ex(t)$ of a Wiener process is identically zero. It is quite often desirable to study a process which has the essential simplicity of the Wiener process but possesses a non vanishing mean value function. We give therefore the following definition:

A stochastic process $y(t)$ is said to be a _Wiener process with mean value function $f(t)$_ if

(2.1) $y(t) = x(t) + f(t).$

Here $f(t)$ is a fixed real valued function of t such that $f(0) = 0$ and $x(t)$ is a Wiener process.

The process $y(t)$ is completely determined by its variance parameter and its mean value function $f(t)$. In order to be able to apply estimation procedures to the function $f(t)$ it is desirable to give it in such a form that it contains certain parameters. A fairly general assumption is that

(2.2) $f(t) = k_1\emptyset_1(t) + k_2\emptyset_2(t)+\ldots \ \ k_s\emptyset_s(t).$

Here the functions $\emptyset_1(t)\ldots\emptyset_s(t)$ are s arbitrary but completely specified functions while the coefficients $k_1\ldots k_s$ are the parameters of the mean value function. The functions $\emptyset_1(t)\ldots\emptyset_s($ are only subject to certain restrictions*[3] which are satisfied in all cases of practical interest.

3. The estimation procedure.

We assume in the following that the $y(t)$ process - as described by (2.1) and (2.2) - is observed over a finite interval $[o,T]$ so that one sample curve is known over this interval.

We compute the following quantities

---

*These assumptions are: (a) All the functions $\emptyset$ (t) are twice differentiable. (b) For any set $a_1,\ldots,a_s$ of real numbers, not all zero, we have $a_1\emptyset_1'(t)+\ldots a_s\emptyset'_s(t) \neq 0$ for some t-set of positive measure which is contained in $[o,T]$.

(3.1) $\quad \Phi_{ij} = \int_0^T \emptyset'_i(t) \, \emptyset'_j(t)dt \quad\quad (i,j = 1,2,\ldots,s)$

and then the quantities Φ^{ij} which are determined by the matrix relation

(3.2) $\quad\quad\quad ((\Phi^{ij})) = ((\Phi_{ij}))^{-1}$

Next one has to compute the integrals

(3.3) $\int_0^T \emptyset'_i(t)dy(t) = y(T)\emptyset'_i(T) - \int_0^T y(t)\emptyset''_i(t)dt \quad\quad (i=1,\ldots,s)$

The estimates $\hat{k}_j$ of the parameters k_j are then

(3.4) $\quad\quad \hat{k}_j = \sum_{v=1}^s \Phi^{jv} \int_0^T \emptyset'_v(t)dy(t) \quad\quad\quad (j=1,2,\ldots,s)$

The estimates $\hat{k}_j$ $(j=1,2,\ldots,s)$ are limits of maximum likelihood estimates computed from observations taken over a finite set of points. It can be shown that under rather general conditions the estimates $\hat{k}_j$ are unbiased estimates of k_j and that the covariance of $\hat{k}_i$ and $\hat{k}j$ is given by $\sigma(\hat{k}_i \hat{k}_j) = c\,\Phi^{ij}$.

It should be emphasized that the estimates $\hat{k}_j$ are not maximum likelihood estimates and that the estimated mean value

(3.5) $\quad\quad \hat{f}(t) = \hat{k}_1 \, \emptyset_1(t) + \ldots + \hat{k}_s \, \emptyset_s(t)$

is not a least square fit of the data. Moreover it can be shown that the mean value curve $\hat{f}(t)$ has another optimal property. We say that $\hat{f}(t)$ is a best linear estimate of $f(t)$ if

(a) $E \hat{f}(t) = f(t)$ (i.e. $\hat{f}(t)$ is an unbiased estimate)

(b) $E \int_0^T [f(t) - \hat{f}(t)]^2 dt \leq E \int_0^T [f(t) - \tilde{f}(t)]^2 dt$

where $\tilde{f}(t)$ is an estimate, $\tilde{f}(t) = \sum_{v=1}^{s} \tilde{k}_v \, \emptyset_v(t)$ such that

$$\tilde{k}_v = \int_0^T \psi(t) dy(t) \text{ and also } E(\tilde{k}_v) = k_v.$$

The estimate (3.5) is a best linear estimate in this sense.

We finally give an estimate for the variance constant.

We divide the interval $[0, T]$ into N equal parts of length $\tau = \frac{T}{N}$ and denote the subdivision points by $t_i = i\tau$ (i = 0,1, 2,...,N). The quantity

(3.6) $\hat{c} = \frac{1}{T - s\tau} \sum_{n=1}^{N} [y(t_n) - y(t_{n-1}) - \hat{f}(t_n) + \hat{f}(t_{n-1})]^2$

can be used as an estimate for the variance constant c. This estimate is slightly biased but its bias converges to zero as N increases.

We still have to choose the functions $\emptyset_1(t)$, $\emptyset_2(t)$,...,$\emptyset_s(t)$. In this report we consider only the case in which the $\emptyset_j(t)$ are polynomials. However other choices are possible. For instance it might be advantageous to choose the $\emptyset_j(t)$ as trigonometric functions when phenomena with a definite periodicity are studied.

As in least squares theory it seems also here to be convenient to use orthogonal polynomials. The purpose of introducing orthogonal polynomials is in the present case the reduction of the matrix $((\phi_{ij}))$ to a diagonal matrix. From (3.1) we see that it is then necessary to assume that the derivatives $\emptyset'_j(t)$ are the first s polynomials of a system orthogonal with respect to the weight function 1 over an interval $[0,T]$. It follows then that the functions $\emptyset_j(t)$ become the integrals of Legendre polynomials adapted to the interval $[0,1]$. As a consequence the mean value function must have initial and terminal value zero, i.e.

$$(3.7) \qquad\qquad f(0) = f(T) = 0 \ .$$

Condition (3.7) will not be fulfilled in general. Moreover, it is not possible to enforce this condition by a rotation of the axis. In the situation under consideration such a rotation would change the character of the process and prevent it from being a Wiener process. This means that in general it will not be possible to use orthogonal polynomials to estimate the mean value function of a Wiener process. The use of orthogonal polynomials is restricted to certain physical situations where the quantity measured starts at a certain level which is reached again at the end of the period of observation.

In the general case (3.7) is not satisfied and it is then convenient to choose for the $\emptyset_j(t)$ consecutive powers of t.

We assume therefore

$$(3.8) \quad \begin{cases} \emptyset_j(t) = t^j & (j = 1,2,\ldots,s) \\ f(t) = k_1 t + k_2 t^2 + \ldots + k_s t^s \end{cases}$$

We obtain therefore from (3.1), (3.3) and (3.4)

(3.9)
$$\Phi_{ij} = \frac{ij}{i+j-1} T^{i+j-1}$$

(3.10)
$$\begin{cases} \int_0^T t^{v-1} dy(t) = T^{i-1} y(T) - (i-1) \int_0^T y(t) t^{i-2} dt \\ \qquad\qquad\qquad\qquad\qquad \text{for } i = 2,\ldots,s \\ \hat{k}_j = \sum_{v=1}^s \Phi^{jv} v \int_0^T t^{v-1} dy(t) \end{cases}$$

It is easy to show that

(3.11)
$$\Phi^{jv} = \frac{S^{jv}}{j \cdot v T^{j+v-1}}$$

where the matrix $\|S^{jv}\|$ is the inverse of the matrix

$$\left\| \frac{1}{j+v-1} \right\|_{j,v=1,\ldots,s} .$$

The computational work involved in this estimation procedure can be greatly reduced by providing tables of the matrices $\|S^{jv}\|$. These are inverses of finite segments of the Hilbert matrix and were tabulated for segments up to and including order 10. These tables [5] should be used whenever a polynomial of degree not exceeding 10 is the mean value function. The scope of these tables will be sufficient in most cases. It should be remarked that with the use of these

tables the estimation of the mean value function is by no means
more laborious than the computation of a least square fit to
the same data. One could even say that the estimation of the
mean value function is in some respects more convenient than
a least square fit. This is due to the fact that the table to
be used for estimating the mean value curve depends only on
the degree of the polynomials but is independent of the
number of points observed in [0,T]. The tables for fitting ORThogon
polynomials by least squares depend on the contrary on the degree of the polynomials as well as on the number of points.

4. Construction of an example.

   We next construct an example to which we will apply
the technique discussed in the preceding section. We use random numbers to construct "data" which simulate observations
from a Wiener process with mean value function.

   We choose the following polynomial of degree four

(4.1) $f(t) = 3400 + 310t - 2.7t^2 + 43 \ 10^{-3}.t^3 - 2.6 \times 10^{-4} t^4$

This polynomial will represent the mean value function of our
fictitious process. We assume that T = 100 and that we take
observations at all integer values $0 \leq n \leq 100$. We compute
therefore the values of f(n), n = 0(1)100. In order to
obtain the simulated observations we add random numbers as
"errors" to the values f(n). These random numbers were
obtained from H. Wold's table [6]. These tables contain random

normal deviates representing a normal population of zero mean
and unit variance. They are arranged in colums of 50 and
the sums $\sum(x)$ of each column is also given. We still must
select the variance constant c. In view of the arrangement of
Wold's tables it was decided to choose

(4.2) $$c = \frac{50}{3} .$$

Random numbers from a normal population with zero mean and
variance $c = \frac{50}{3}$ ⌃obtained by dividing the values $\sum(x)$ in
were
Wold's table by $\sqrt{3}$. A set of 100 random numbers (denoted
in the following by w_i) was derived in this manner by taking
100 columns in Wold's table. We started with the first column
and used all consecutive columns with the exception of the
random numbers on page 7. These were skipped in agreement
with the warning given in the tables. The sum $\sum_{i=1}^{n} w_i$
(n = 0,1,2,...,100) with $w_0 = 0$ were computed and added as
"errors" to the "true values". In this manner

(4.3) $$g(\blacksquare) = f(\blacksquare) + \sum_{i=0}^{n} w_i \qquad (i = 0,1,...,100)$$

was computed. These values of $g(\blacksquare)$ are the simulated observations. We will apply in the following our estimation procedure to these simulated observations. Since we wish to
use a Wiener process as our model we have to adjust the observations to make the initial value equal to zero. This can

complished by subtracting g(0) from all observations.

leaves the increments unchanged and does therefore not

t the character of the process. We denote by

$$y(n) = g(n) - g(0).$$

1 gives for each n the values of $f(n)$, $\sum_{i=1}^{n} w_i$, $g(n)$

(n).

n	f(n)	$\sum_{i=0}^{n} w_i$	g(n)	y(n)
0	3400.00	0.00	3400.00	0.00
1	3707.34	+6.05	3713.39	313.39
2	4009.54	+7.02	4016.56	616.56
3	4306.84	+2.56	4309.40	909.40
4	4599.49	−0.41	4599.08	1199.08
5	4887.71	−3.74	4883.97	1483.97
6	5171.75	−11.26	5160.49	1760.49
7	5451.82	−14.59	5437.23	2037.23
8	5728.15	−11.66	5716.49	2316.49
9	6000.94	−13.71	5987.23	2587.23
10	6270.40	−13.70	6256.70	2856.70
11	6536.73	−12.81	6523.92	3123.92
12	6800.11	−12.26	6787.85	3387.85
13	7060.75	−13.22	7047.53	3647.53
14	7318.80	− 4.09	7314.71	3914.71
15	7574.46	− 5.67	7568.79	4168.79
16	7827.89	− 6.09	7821.80	4421.80
17	8079.24	− 2.63	8076.61	4676.61
18	8328.68	− 3.20	8325.48	4925.48
19	8576.35	− 0.32	8576.03	5176.03
20	8822.40	− 0.21	8822.19	5422.19
21	9066.96	+ 7.93	9074.89	5674.89
22	9310.16	+ 9.19	9319.35	5919.35
23	9552.12	+ 8.69	9560.81	6160.81
24	9792.97	+ 4.34	9797.31	6397.31
25	10032.81	+ 5.51	10038.32	6638.32
26	10271.75	+ 6.33	10278.08	6878.08
27	10509.89	+ 6.44	10516.33	7116.33
28	10747.33	+13.89	10761.22	7361.22
29	10984.13	+18.07	11002.20	7602.20
30	11220.40	+28.26	11248.66	7848.66
31	11456.19	+27.28	11483.47	8083.47
32	11691.59	+22.54	11714.13	8314.13
33	11926.65	+22.92	11949.57	8549.57
34	12161.42	+29.42	12190.84	8790.84
35	12395.96	+30.64	12426.60	9026.60
36	12630.31	+33.62	12663.93	9263.93
37	12864.50	+30.43	12894.93	9494.93
38	13098.56	+31.53	13130.09	9730.09
39	13332.52	+22.28	13354.80	9954.80
40	13566.40	+19.28	13585.68	10185.68

n	f(n)	$\sum_{i=0}^{n} w_i$	g(n)	y(n)
40	13566.40	19.28	13585.68	10185.68
41	13800.20	24.97	13825.17	10425.17
42	14033.94	24.01	14057.95	10757.95
43	14267.61	34.91	14302.52	10902.52
44	14501.21	39.54	14540.75	11140.75
45	14734.71	37.63	14772.34	11372.34
46	14968.11	39.20	15007.31	11607.31
47	15201.37	38.24	15239.61	11839.61
48	15434.48	41.34	15475.82	12075.82
49	15667.38	43.40	15710.77	12310.77
50	15900.00	41.94	15941.94	12541.94
51	16132.34	39.91	16172.25	12772.25
52	16364.32	46.44	16410.76	13010.76
53	16595.89	54.57	16650.46	13250.46
54	16826.96	51.10	16878.06	13478.06
55	17057.46	44.44	17101.90	13701.90
56	17287.32	47.63	17334.95	13934.95
57	17516.44	48.87	17565.31	14165.31
58	17744.73	51.55	17796.28	14396.28
59	17972.07	55.07	18027.14	14627.14
60	18198.40	55.15	18253.55	14853.55
61	18423.56	56.79	18480.35	15080.35
62	18647.46	58.08	18705.54	15305.54
63	18869.95	64.22	18934.17	15534.17
64	19090.92	65.90	19156.81	15756.81
65	19310.21	64.83	19375.04	15975.04
66	19527.70	65.97	19593.67	16193.67
67	19743.22	65.96	19809.18	16409.18
68	19956.62	65.29	20021.91	16621.91
69	20167.74	67.54	20235.28	16835.28
70	20376.40	74.56	20450.96	17050.96
71	20582.44	78.49	20660.93	17260.93
72	20785.66	78.65	20864.31	17464.31
73	20985.89	79.59	21065.48	17665.48
74	21182.92	81.32	21264.24	17864.24
75	21376.56	83.12	21459.68	18059.68
76	21566.60	84.18	21650.78	18250.78
77	21752.83	84.83	21837.66	18437.66
78	21935.02	81.67	22016.69	18616.69
79	22112.96	86.44	22199.40	18799.40
80	22286.40	82.16	22368.56	18968.56

n	f(n)	$\sum_{i=0}^{n} w_i$	g(n)	y(n)
80	22286.40	82.16	22368.56	18968.56
81	22455.12	81.37	22536.49	19136.49
82	22618.86	80.44	22699.30	19299.30
83	22777.38	84.07	22861.45	19461.45
84	22930.42	82.34	23012.76	19612.76
85	23077.71	82.09	13159.80	19759.80
86	23219.00	78.16	23297.16	19897.16
87	23353.99	73.10	23427.09	20027.09
88	23482.42	74.72	23557.14	20157.14
89	23603.98	75.86	23679.84	20279.84
90	23718.40	84.38	23802.78	20402.78
91	23825.36	86.21	23911.57	20511.57
92	23924.57	86.97	24011.54	20611.54
93	24015.70	91.86	24107.56	10707.56
94	24098.44	90.60	24189.04	20789.04
95	24172.46	81.73	24254.19	20854.19
96	24237.44	78.38	24315.82	20915.82
97	24293.03	80.21	24373.24	20973.24
98	24338.88	80.01	24418.89	21018.89
99	24374.66	83.05	24457.71	21057.71
100	24400.00	79.30	24479.30	21079.30

5. Computation of the estimated mean value curve and of the variance constant

We next compute the estimate of the mean value function using formulae (3.9), (3.10) and (3.11) in order to obtain the integral (3.10) we have to compute the integrals $\int_0^T y(t)\, t^{i-2} dt$ for $i = 2,\ldots,s$. These integrals were evaluated by means of Simpson's rule

$$(5.1) \qquad \int_{x_0}^{x_0+nh} y\, dx = \frac{h}{3}[y_0 + 4(y_1 + y_3 + \ldots + y_{n-1}) + 2(y_2 + y_4 + \ldots + y_{n-2})$$
$$+ y_n]$$

where n is an even number.

The results were compared in all cases to the approximation by summation and for $i = 5$ also to the use of a seven point Lagrangian interpolation formula [1]. It was found that results obtained by the seven-point formula and Simpson's rule differ only little while straight summation would be too crude an approximation to the integral. As an example, we mention that $\int_0^T y(t) t^3 dt = 461,157,455,032$ by Simpson's rule, while $461,160,141,829$ was obtained from the 7-point formula and $471,752,177,029$ by straight summation. The integrals $\int_0^T y(t) t^{i-2} dt$ were therefore computed by (5.1), the values are given in table 2.

Table 2

$$i \qquad \int_0^{T} y(t)t^{i-2}\, dt$$

i	
2	1,209,256.51
3	78,807,137.03
4	5,825,790,539.54
5	461,157,455,032.19
6	38,111,464,889,820.59

We next compute from table 2 the values $a_i = iy(T)T^{2-1} - i(i-1)\int_0^T y(t)t^{i-2}dt$ for $i = 2,\ldots,6$ while $a_1 = y(T) = y_{100}$. These are given in table 3.

Table 3

i	a_i
1	2.10793×10^4
2	1.79735×10^6
3	1.59536×10^8
4	1.44077×10^{10}
5	1.31650×10^{12}
6	1.21414×10^{14}

We finally compute form (3.10) the estimates $\hat{k}_j$ of the coefficients of the mean value function and obtain

$$\hat{k}_j = \sum_{v=1}^{s} \phi^{jv} a_v \text{ for } j = 1,\ldots,s.$$ The elements ϕ^{jv}

of the inverse matrix were taken from the tables [5] mentioned above. The coefficients $\hat{k}_j$ were computed for mean value function of various degrees, using always the same "data" to be fitted. The results of these computations

Table 4

timating mean value curves of degree s

	$s=5$	$s=6$
73	307.495	300.785
211	-2.4932	-1.4868
166x10^{-2}	3.8332x10^{-2}	-8.638x10^{-3}
753x10^{-4}	-2.128x10^{-4}	7.266x10^{-4}
	-1.79x10^{-7}	-8.63352x10^{-6}
		2.8182x10^{-8}

ts the values $\hat{f}_s(t)$ of the estimated

gree s were computed for t=1 and

4,5,6. In addition the mean value es-

4 was computed for t=0(1)100 and a

e simulated observations was also de-

n of these curves given in table 5,

to compare the 4-th degree estimate with

squares fit. We denote by $\hat{f}_4(t)$ the

estimate and by $\tilde{f}_4(t)$ the 4-th degree

we write again y(t) for the (adjusted)

then we see from table 6 that

t)-y(t)$]^2$ = 3351.71

t)-y(t)$]^2$ = 3166.90

Table 5

Comparison between the coordinates, the observations and
the true values

t	"true values" f(t)-f(0)	"Simulated observations" y(t)	Least squares fit f₄(t)
1	307.34	313.39	303.52
10	2870.40	2856.70	2865.12
20	5422.40	5422.19	5423.79
30	7820.40	7848.66	7834.18
40	10166.40	10185.68	10195.57
50	12500.00	12541.94	12545.23
60	14798.40	14853.55	14858.43
70	16976.40	17050.96	17048.42
80	18886.40	18968.56	18966.44
90	20318.40	20402.78	20401.75
100	21000.00	21079.30	21081.58

Estimates of mean value curve of degree

t	3	4	5	6
1	257.73	306.19	305.04	299.29
10	2610.26	2865.21	2861.83	2991.90
20	5258.32	5423.14	5424.66	5442.31
30	7888.16	7834.08	7839.22	7851.25
40	10443.74	10196.58	10200.83	10191.15
50	12869.00	12547.08	12547.31	12526.12
60	15107.92	14860.69	14856.81	14846.98
70	17104.44	17050.36	17045.67	17057.57
80	18802.54	18967.36	18966.27	18983.58
90	20146.18	20401.13	20404.87	20399.69
100	21082.70	21079.30	21079.50	21079.30

Table 6

t	y(t)	$\hat{f}_4(t)$	$\tilde{f}_4(t)$
0	0	0	-3.11
1	313.39	306.19	303.52
2	616.56	607.39	605.13
3	909.40	903.85	901.95
4	1199.08	1195.79	1194.23
5	1483.97	1483.45	1482.21
6	1760.49	1767.05	1766.09
7	2037.23	2046.82	2046.12
8	2316.49	2322.96	2322.50
9	2587.23	2595.70	2595.43
10	2856.70	2865.21	2865.12
11	3123.92	3131.70	3131.77
12	3387.85	3395.36	3395.57
13	3647.53	3656.37	3656.69
14	3914.71	3914.90	3915.32
15	4168.79	4171.12	4171.62
16	4421.80	4425.20	4425.76
17	4676.61	4677.30	4677.90
18	4925.48	4927.56	4928.19
19	5176.03	5176.13	5176.77
20	5422.19	5423.14	5423.79
21	5674.89	5668.74	5669.38
22	5919.35	5913.05	5913.66
23	6160.81	6156.18	6156.76
24	6397.31	6398.27	6398.79
25	6638.32	6639.40	6639.86
26	6878.08	6879.67	6880.08
27	7116.33	7119.20	7119.53
28	7361.22	7358.06	7358.31
29	7602.20	7596.33	7596.50
30	7848.66	7834.08	7834.18
31	8083.47	8071.42	8071.43
32	8314.13	8308.42	8308.30
33	8549.57	8545.08	8544.85
34	8790.84	8781.48	8781.14
35	9026.60	9017.66	9017.22
36	9263.93	9253.68	9253.12
37	9494.93	9489.55	9488.88
38	9730.09	9725.32	9724.53
39	9954.80	9960.99	9960.09
40	10185.68	10196.58	10195.57

	$y(t)$	$f_4(t)$	$f_4(t)$
40	10185.68	10196.58	10195.57
41	10425.17	10432.11	10430.99
42	10657.95	10667.58	10666.34
43	10902.52	10902.98	10901.63
44	11140.75	11138.30	11136.84
45	11372.34	11373.52	11371.96
46	11607.31	11608.64	11606.97
47	11839.61	11843.60	11841.84
48	12075.82	12078.39	12076.54
49	12310.77	12312.97	12311.02
50	12541.94	12547.09	12545.23
51	12772.25	12781.06	12779.14
52	13010.76	13014.66	13012.67
53	13250.46	13247.81	13245.77
54	13478.06	13480.45	13478.36
55	13701.90	13712.50	13710.36
56	13934.95	13943.86	13941.69
57	14165.31	14174.47	14172.26
58	14396.28	14404.21	14401.98
59	14627.14	14632.98	14630.74
60	14853.55	14860.69	14858.43
61	15080.35	15087.20	15084.94
62	15305.54	15312.41	15310.20
63	15534.17	15536.17	15533.93
64	15756.81	15758.37	15756.15
65	15975.04	15978.86	15976.67
66	16193.67	16197.50	16195.34
67	16409.18	16414.13	16412.01
68	16621.91	16628.59	16626.52
69	16835.28	16840.73	16838.72
70	17050.96	17050.36	17048.42
71	17260.93	17257.32	17255.45
72	17464.31	17461.42	17459.63
73	17665.48	17662.48	17660.77
74	17864.24	17860.29	17858.68
75	18059.68	18054.66	18053.14
76	18250.78	18245.38	18243.97
77	18437.66	18432.23	18430.93
78	18616.69	18614.99	18613.82
79	18799.40	18793.45	18792.40
80	18968.56	18967.36	18966.44

t	y(t)	$f_4(t)$	$f_4(t)$
80	18968.56	18967.36	18966.44
81	19136.49	19136.50	19135.71
82	19299.30	19300.60	19299.96
83	19461.45	19459.43	19458.93
84	19612.76	19612.73	19612.38
85	19759.80	19760.23	19760.03
86	19897.16	19901.66	19901.62
87	20027.09	20036.75	20036.93
88	20157.14	20165.22	20165.51
89	20279.84	20286.78	20287.23
90	20402.78	20401.13	20401.75
91	20511.57	20507.98	20508.77
92	20611.54	20607.02	20607.98
93	20707.56	20697.93	20699.06
94	20789.04	20780.40	20781.70
95	20854.19	20854.10	20855.57
96	20915.82	20918.71	20920.34
97	20973.24	20973.88	20975.68
98	21018.89	21019.26	21021.23
99	21057.71	21054.52	21056.65
100	21079.30	21079.30	21081.58

We still have to estimate the variance constant. This is done for the estimate $\hat{f}_4(t)$ by means of (3.6) and the data contained in table 6. We obtain the estimate $\hat{c} = 15.051$ as compared with the "true value" $c = 50/3$.

We also give four figures which permit to visualize the relative errors of our estimates. Figures 1 and 2 compare the estimating polynomials of degrees 3, 4, 5 and 6. In figure 1 the graphs of

$$\hat{R}_4(t) = \frac{\hat{f}_4(t) - f(t)}{f(t)} \quad \text{and of} \quad \tilde{R}_4(t) = \frac{\tilde{f}_4(t) - f(t)}{f(t)}$$

are given, in figure 2 we find the corresponding errors

referred to the observations, i.e. the curves $\hat{r}_4(t) =$

$$= \frac{\hat{f}_4(t) - y(t)}{y(t)} \quad \text{and} \quad \tilde{r}_4(t) = \frac{\tilde{f}_4(t) - y(t)}{y(t)} \quad .$$

In figures 3 and 4 we compare the error of the estimating polynomials of degress 3,4,5 and 6, the basis of reference is in figure 3 the curve $f(t)$ in figure 4 the curve $y(t)$.

6. Conclusion. In actual applications we will often encounter situations where more than a single sample curve of the process are given. In these cases a number of problems can arise. It might for instance be necessary to test the hypothesis whether two or more sample curves come from processes with the same variance constant or with the same mean value function. Due to the fact that increments from a Wiener process behave as if they were independent observations from a normal population many of the questions which might arise can be treated by standard statistical techniques. To go into such a discussion would however exceed the scope of this paper.

The author wishes to acknowledge his thanks to Mr. Edwin L. Grab who carried out the necessary computations.

References

] G. Blanch, I. Rhodes, Seven-point Lagrangian integration formulas. Jour. Math. Phys. 22, 204-207 (1943).

] H. B. Mann, The estimation of parameters in certain stochastic processes. Sankhya 11, 97-106 (1951).

] H. B. Mann, On the estimation of parameter determining the mean value function of a stochastic process. To be published in Sankhya.

] Statistical Techniques Applicable to the analysis of a fundamental random process. Technical Note W.C.R.R-52-7. Flyhd. Res. Lab., Wright Air Dev. Center, Dayton, Ohio, 1952.

] I. R. Savage, E. Lukacs, Tables of inverses of finite segments of the Hilbert Matrix. To be published in the NBS Applied Math. Series.

] H. Wold. Random Normal Deviates. Tracts for Computers XXV. Cambridge University Press, 1948.

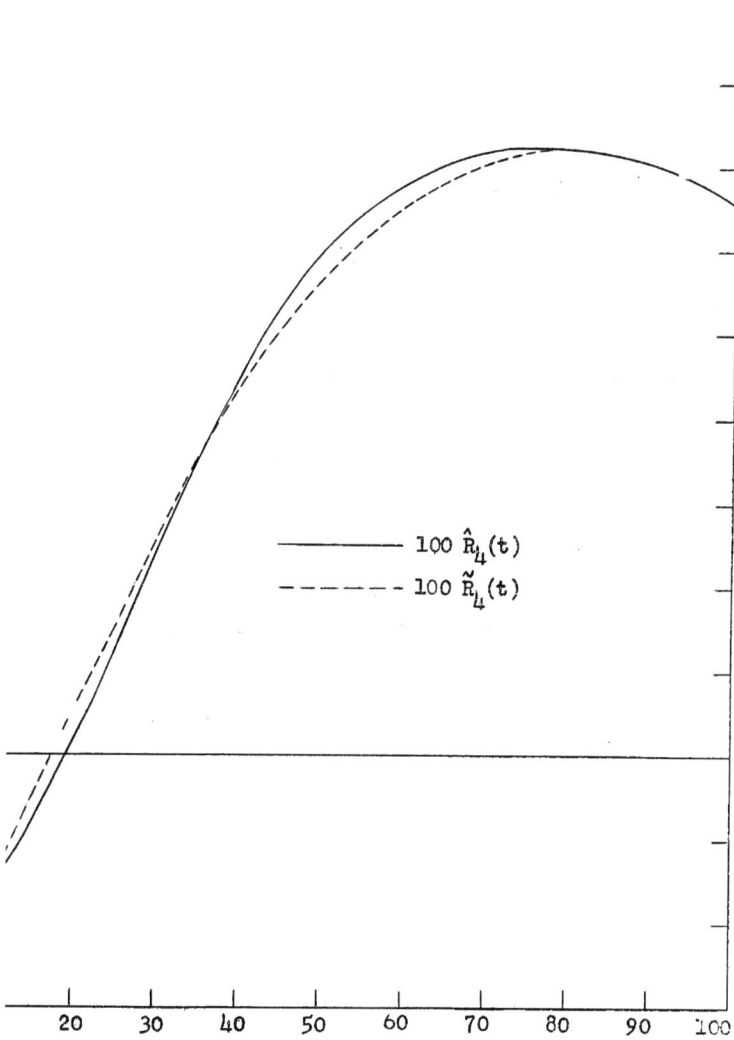

FIG. 1

$100 \ \hat{R}_4(t)$ and $100 \ \tilde{R}_4(t)$

——— $100 \ \hat{R}_4(t)$

– – – $100 \ \tilde{R}_4(t)$

20   30   40   50   60   70   80   90   100

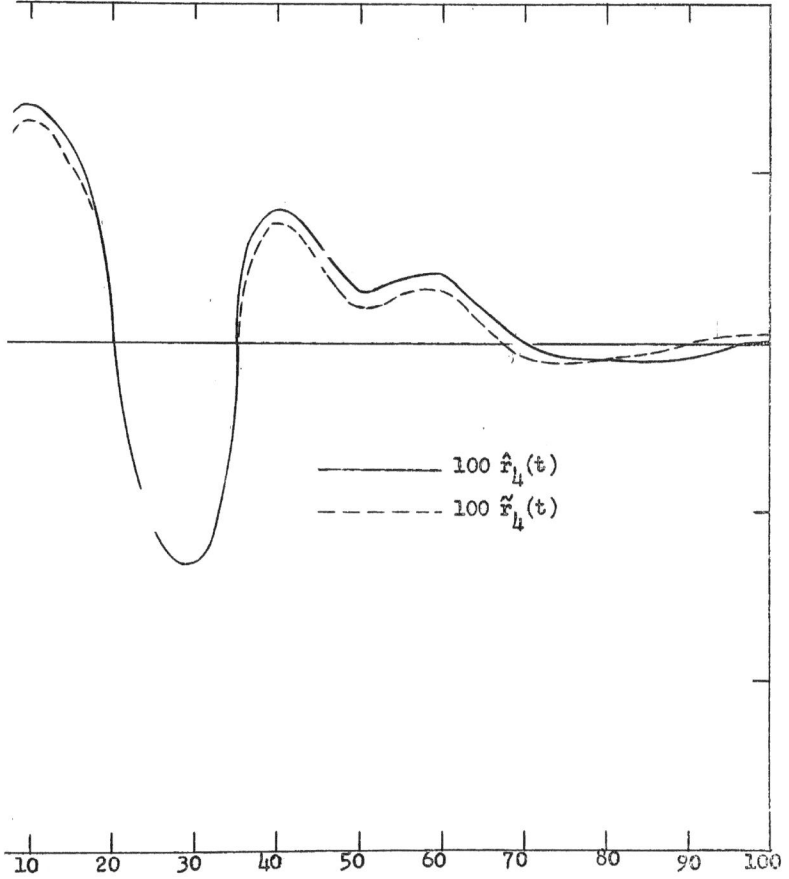

FIG. 2

$100 \; \hat{r}_4(t)$ and $100 \; \tilde{r}_4(t)$

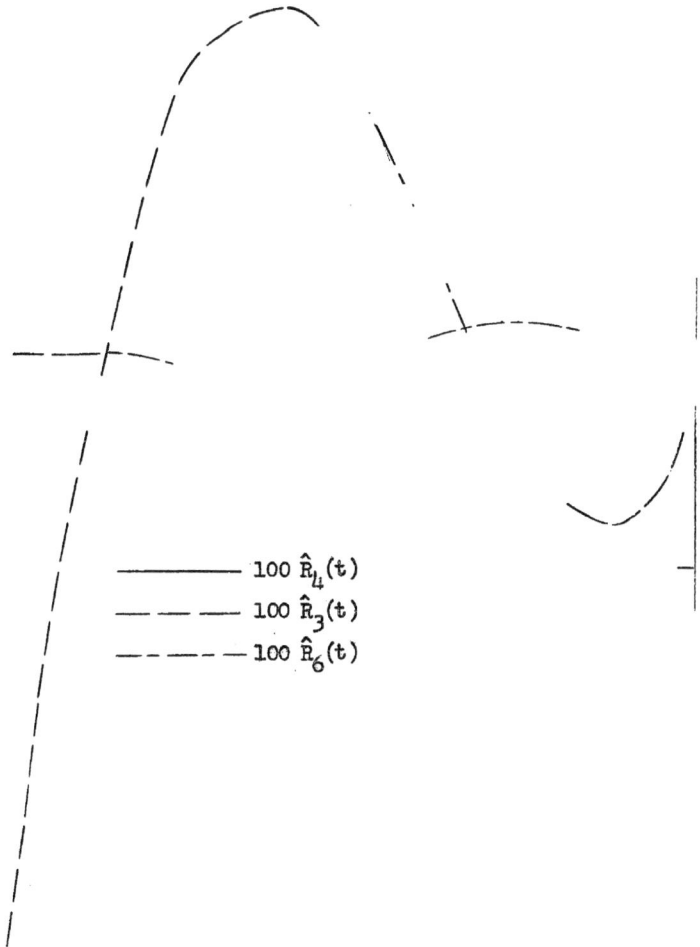

FIG. 3

$100 \ \hat{R}_j(t)$

100 $\hat{R}_4(t)$
100 $\hat{R}_3(t)$
100 $\hat{R}_6(t)$

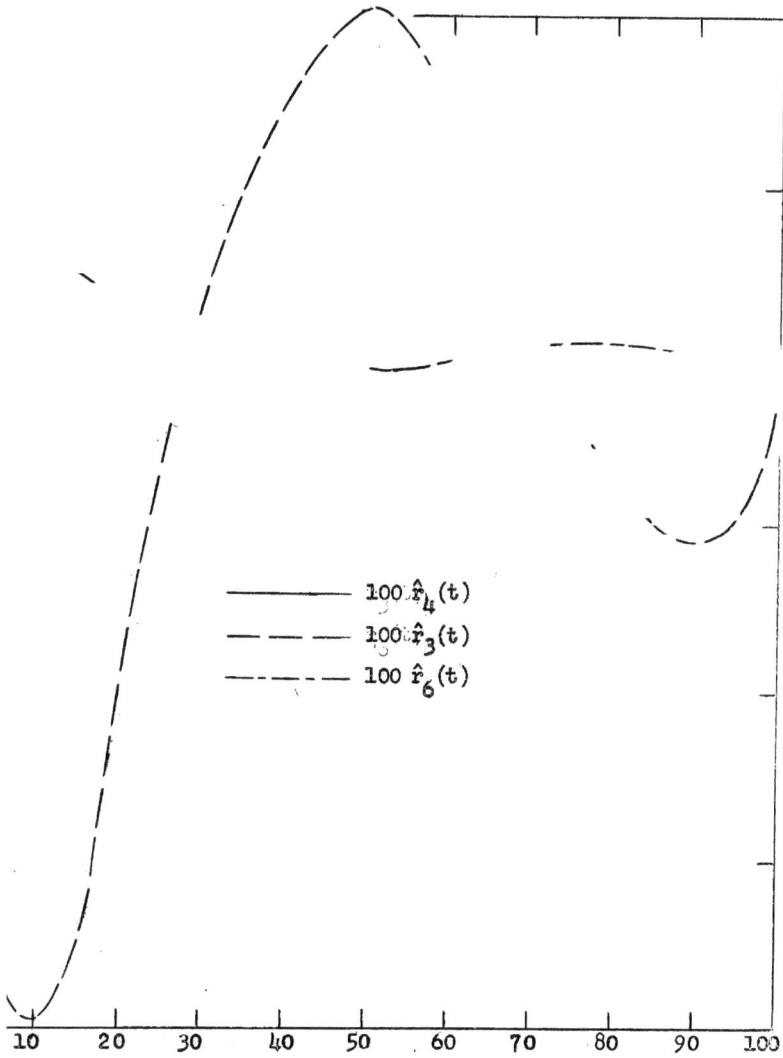

FIG. 4

$100 \, \hat{r}_j(t)$

———	$100 \, \hat{r}_4(t)$	
— — —	$100 \, \hat{r}_3(t)$	
—·—·—	$100 \, \hat{r}_6(t)$	

THE NATIONAL BUREAU OF STANDARDS

Activities

ns of the National Bureau of Standards are set forth in the Act of Congress, March
aded by Congress in Public Law 619, 1950. These include the development and
the national standards of measurement and the provision of means and methods
surements consistent with these standards; the determination of physical constants
of materials; the development of methods and instruments for testing materials,
ructures; advisory services to Government Agencies on scientific and technical
tion and development of devices to serve special needs of the Government; and the
standard practices, codes, and specifications. The work includes basic and applied
pment, engineering, instrumentation, testing, evaluation, calibration services, and
tion and information services. A major portion of the Bureau's work is performed
ment Agencies, particularly the Department of Defense and the Atomic Energy
he scope of activities is suggested by the listing of divisions and sections on the
it cover.

ublications

of the Bureau's work take the form of either actual equipment and devices or
s and reports. Reports are issued to the sponsoring agency of a particular project
ablished papers appear either in the Bureau's own series of publications or in the
essional and scientific societies. The Bureau itself publishes three monthly peri-
e from the Government Printing Office: The Journal of Research, which presents
reporting technical investigations; the Technical News Bulletin, which presents
reliminary reports on work in progress; and Basic Radio Propagation Predictions,
lata for determining the best frequencies to use for radio communications throughout
ere are also five series of nonperiodical publications: The Applied Mathematics
s, Handbooks, Building Materials and Structures Reports, and Miscellaneous

on the Bureau's publications can be found in NBS Circular 460, Publications of
ureau of Standards ($1.00). Information on calibration services and fees can be
Circular 483, Testing by the National Bureau of Standards (25 cents). Both are
the Government Printing Office. Inquiries regarding the Bureau's reports and
uld be addressed to the Office of Scientific Publications, National Bureau of Stand-
n 25, D. C.

:

CPSIA information can be obtained
at www.ICGtesting.com
Printed in the USA
BVHW041521231118
533817BV00018B/1074/P